Earthworks

P. C. Horner

Thomas Telford Ltd
1981

CONTENTS

Other ICE Works Construction Guides available
Pile Driving, W. A. Dawson
Access scaffolding, C. J. Wilshere

ISBN 0 7277 0091 X

Published by Thomas Telford Ltd, Telford House, PO Box 101, 26–34 Old Street, London EC1P 1JH

Typeset, printed and bound by David Green (Printers) Ltd, Kettering, Northamptonshire

1. INTRODUCTION

The process whereby the surface of the earth is excavated and transported to and compacted at another location, is collectively called Earthworks.

Man has been reshaping his environment since prehistoric times in order to improve the conditions of his society; the earliest works were probably the excavation of caves. Many of the older recorded large schemes were carried out in the Middle East and Far East, such as the basin irrigation schemes of King Menes in about 3100 BC and the kanats (water supply tunnels) of Iran that date back to 2800 BC.

Earthmoving continued to be carried out largely by manual and simple mechanical means until the development of steam powered excavating plant in the middle of the nineteenth century. Since then the introduction of the internal combustion engine, electric power, and more recently hydraulic power, have led to the development of a wider range of earthmoving and compaction plant. In addition there has been a general increase in the size and capacity of the plant to meet the demands of increased productivity. The use of modern plant is general in the UK and in other areas where western technology has been adopted; however, in under-developed countries such as China, and parts of Africa, manual labour is still extensively used. The fastest rate of excavation on record was achieved by manual labour without mechanical plant on the North Kiangsu Canal in China where 70 million cubic metres of earthworks were constructed in eighty days.

In the UK the scale of earthworks carried out ranges from small works, such as the excavation of ditches and trenches for drainage and pits for foundations, to large earthworks required for the construction of motorways and dams. Earthworks are frequently carried out at an early stage in a construction project, often as a precursor to the more expensive above ground construction. Completion of the earthworks within the prescribed programme time is often the key to the completion on time of the whole project. Success depends on choosing the correct types and sizes of machine to meet the particular requirements of the site, and then using them in the most effective way.

2. THE DESIGN OF EARTHWORKS

Design of cuttings and embankments

There are three main aspects of earthworks: the clearance of land of vegetation and existing structures; the excavation of ground to form trenches, pits and cuttings; the construction of embankments or fills. Land clearance and demolition may use standard earthmoving or specialist plant and is discussed further in section 3 (Earthmoving practice).

Excavations

Cuttings may be formed either as part of the permanent works or as a temporary expedient in the construction of the works. Permanent cuttings are generally required to be stable to an acceptable degree during their proposed lifetime, whilst temporary cuttings require to be stable only for the time taken to construct the structure for which they have been excavated (local failures may be tolerable.)

Stability of cuttings can generally be achieved by excavating the material to stable slope batters or by retaining the material. The stability of natural and cut slopes is controlled largely by the density and strength of the materials that form the slope over the time period required, the groundwater conditions or porewater pressures, and the strength and disposition of any discontinuities; a factor of safety of at least 1.3 is usually applied. It is generally recommended that specialist geotechnical advice be obtained when designing such slopes. In Table 1, typical slope batters are given for both the temporary (short term) and permanent (long term) cases. It is evident that in cohesive soils the slope angles required for stability are steeper for the short term case than for the long term case, as the strength of cohesive soils is usually greater in the short term. However, it should be noted that where a cutting exists for some time the stability will change from the short term to long term condition, and in addition weathering of the slopes may occur and these factors should be allowed for in design.

Table 1. Typical batters of excavated slopes

	Slope batters* (vert : horiz)					
Material	Permanent			Temporary		
Massive rock	1½:1	to	vertical	1½:1	to	vertical
Well jointed/bedded rock	1:2	to	2:1	1:2	to	2:1
Gravel	1:2	to	1:1	1:2	to	1:1
Sand	1:2½	to	1:1½	1:2½	to	1:1
Clay	1:6	to	1:2	1:2	to	2:1

*These batters are given as a guide and should not be taken for final design purposes.

Retaining structures are generally constructed of timber or steel for temporary works and steel and concrete for permanent works. They should be designed to withstand the pressures generated by the in situ and backfill materials, surcharge loadings and the ambient groundwater conditions. These are controlled by factors similar to those for the stability of slope batters, although a factor of safety of at least 2 is usually applied. The presence of groundwater tends to reduce the stability of slopes and retaining structures and in addition may render the material unsuitable or troublesome to use as fill within the works. In such cases temporary or permanent drainage and dewatering measures may be adopted.

Further information on the design of slopes and retaining structures can be obtained from numerous texts, some of which are included in the bibliography.

Embankments

Embankments may be constructed from a wide range of materials and for numerous purposes. Load-bearing fills, such as those constructed under roads and structures and as part of water retaining structures, require a higher factor of safety than non-load bearing fills such as stockpiles or landscaping fills. Nevertheless, a number of factors should be considered in order that an embankment can be designed and constructed to satisfactorily carry out its function. It is frequently necessary to be selective in the choice of materials to be used in construction in order that the embankment can be constructed economically, and this is discussed later. The major factors to be considered in the design of an embankment are

(a) the stability of the embankment
(b) the bearing capacity of the embankment
(c) the settlement of the embankment
(d) the trafficability of the embankment.

The stability of an embankment may vary with time and this is generally most marked where it is constructed of cohesive soil. Embankments should be analysed for stability in both the long and short term for the ranges of constructed height and materials that are likely to be used in its construction. The stability of the embankment foundations should also be investigated especially when the embankment is constructed on side-long ground or on weak soils. Embankments are generally designed to be stable even when the worst conditions that may reasonably occur during their anticipated lifetimes are taken into account. Alternatively, if excessively low slope batters are found to be necessary with the on-site materials it may be necessary to import material of higher strength (for example rockfill), to improve the strength of the on-site materials (by drying them out prior to use or by chemically stabilizing them with lime, cement, etc.), to emplace drainage measures to dewater the embankment, or partially or totally retaining the fill. Typical slope angles that have been used in the construction of embankments are included in Table 2.

The bearing capacity of an embankment will only be important where the surface is to be loaded with, for example, a road, railway or foundations, and the bearing capacity should be determined for the moisture conditions that are likely to occur during its operation after the equilibrium moisture content has been achieved. For roads, railways and runways the California Bearing Ratio (CBR) or modulus of subgrade reaction are usually used in the design of the pavement, whereas for foundations plate bearing or triaxial test data are often used. Where the bearing capacity of the fill is likely to be inadequate, better quality fill such as gravel or rockfill may be used in the upper layers, the material stabilized chemically or improved with techniques such as dynamic consolidation.

Table 2. Typical batters of embankment slopes

Material	Typical slope batters* (vert : horiz)
Hard rock fill	1:1½ to 1:1
Weak rock fill	1:2 to 1:1¼
Gravel	1:2 to 1:1¼
Sand	1:2½ to 1:1½
Clay	1:2½ to 1:1½

*These batters are given as a guide and should not be taken for final design purposes.

Settlement of embankments may arise from settlement of the fill itself or of the underlying foundation strata. Settlements that occur during the construction of the earthworks can be minimized by providing compaction to the fill, and compensated for by laying extra fill. Differential settlement often occurs where earthworks meet structures because the backfill immediately behind the structures does not generally have as much time to settle as the fill in the embankments and also because the degree of compaction that can be afforded is often less. The settlement of the foundation strata can be similarily compensated for by the addition of fill where the time taken to consolidate occurs within the earthworks programme. If longer periods are required for consolidation, and the long term settlement cannot be tolerated, it may be necessary to provide drainage to facilitate consolidation or to induce it by techniques such as dynamic consolidation, or by surcharging the fill.

The trafficability of potential fill materials is frequently the limiting factor of the efficient construction of embankments, especially with cohesive or saturated granular soils. In practice it has been found that materials which could be used to form stable embankments have been classified as unsuitable, because the specification defining suitability has been written to permit the use of heavy, large, rubber-tyred scrapers, and not lighter plant which could work on poorer quality materials which would nevertheless form stable embankments.

Selection of earthworks materials

In order that an embankment can be constructed economically, and possess adequate stability and acceptable settlement, it is frequently necessary to be selective in the material to be placed in the embankment. Therefore, the classification of the earthworks materials into those that are 'suitable' and 'unsuitable' for incorporation into the works is often made at the design, planning and construction stages of a project. The factors that should be taken into account when choosing suitability criteria can be obtained from a consideration of

(a) the stability of the embankment

(b) the required settlement characteristics of the embankment

(c) the trafficability by earthmoving plant

(d) economic factors.

In general, the parameters taken to define the suitability of cohesive materials are the moisture content, the liquid and plastic (Atterberg) limits, the strength and compaction characteristics. With the development of the Method Specification for compaction the use of the compaction characteristics has become less popular, as it involves more detailed testing at both the design and construction stages. The most common means of defining suitability in the UK is to relate the moisture content to the plastic limit, and typically the ranges of acceptable moisture content lie from the plastic limit minus 4, to 1.2 times the plastic limit.

This means of specifying suitability is less than satisfactory, as the plastic limit test takes some time to carry out and has a poor degree of accuracy. Neither is it a reliable indicator of the strength of the material, which controls the stability of the embankment and the trafficability of plant. The recognition of this, and the fact that these specifications have resulted in the inefficient use of materials, have led to the present tendency to define suitability of cohesive soils on the basis of maximum moisture content and minimum strength criteria. On some recent motorway schemes a minimum strength of about 35 kN/m^2 has been specified (i.e. that required for the trafficability of a large dozer). Where cohesive soils contain a high proportion of gravel, an (upwards) adjustment of the moisture content of the clay to sand matrix may be called for to compensate for the low moisture of the gravel fraction.

It is common practice to specify the suitability of granular soils on the basis of maximum and minimum moisture contents that are related to the optimum moisture content, to ensure that adequate compaction is achieved. Strength criteria, such as the minimum acceptable angle of friction, Φ, or CBR, may also be specified.

There are a number of other criteria which are commonly accepted. Thus, highly organic soils such as peat and timber, and material susceptible to spontaneous combustion, are generally regarded as 'unsuitable', as are frozen materials and highly plastic clays. The criteria for 'suitability' can be altered to suit the economic requirements of the scheme. For example, where an excess of 'suitable' materials are present, a stringent specification may be adopted to ensure that the better materials are used in construction; or, where material may otherwise have to be imported, it may be possible to restrict the method of working and/or design the works to permit the use of materials that would otherwise be unsuitable.

Excavability of materials

The excavability of material is the relative ease or otherwise by which it can be excavated and is related to the properties of the material and the type and size of the available excavating plant. Most excavators dig by pulling, pushing or driving a blade or cutting device into the material to loosen or pick it up. The resistance to penetration of the blade must therefore be overcome before excavation can be achieved. This resistance is initially provided by the shear strength of the material, taken as the cohesion, frictional strength or cemented strength of the material. Cobbles, boulders or other masses tend to increase this resistance, whilst discontinuities such as bedding planes, joints, faults, etc., tend to reduce the resistance. As the digging device penetrates the ground it is further resisted by the adhesion or frictional resistance between the sides of the device and the material and by the weight of the material being excavated.

Materials can be classified into three broad groups on the basis of their excavability, i.e. materials that can be excavated by

(a) easy digging – such as very soft to firm clays; very loose to medium dense sands and gravels

(b) hard digging – such as stiff to hard clays; dense to very dense sands and gravels; weak rocks such as rock chalk; shale and slate and weathered harder rocks

(c) rock – such as limestone, granite and dolerite.

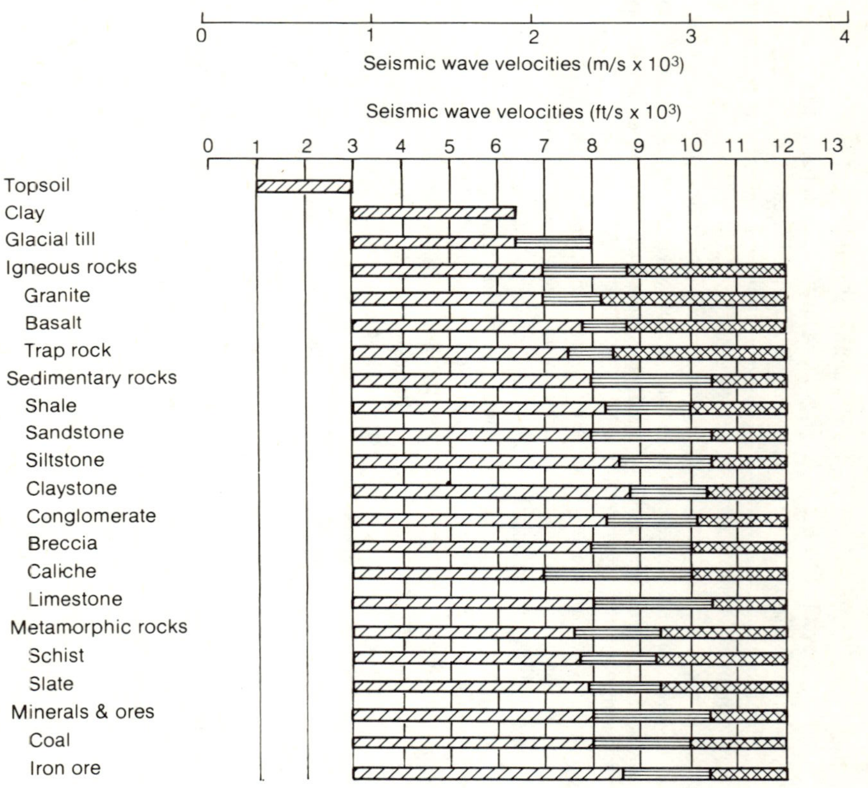

Fig. 1. Ripper performance chart for Caterpillar D9H dozer with multi/single shank 9D ripper (Reproduced by courtesy of Caterpillar Overseas S.A.)

The relative ease of excavation is also related to the utilizable power of the plant being used; within limits, the higher the power of the plant, the greater is the resistance that can be overcome. A further consideration on the ability of plant to excavate material is the location and accessibility of the material. Thus, a motor scraper excavates material from below itself and requires the material to be of sufficient extent, of a limited range of strength and lying at a low gradient; back-actors operate best in removing material below the level of their tracks and can operate on ground of lower strength than motor scrapers. Forward loaders and face shovels generally excavate material above the level of their tracks and a dragline can excavate materials at distances of 20 m or so from the machine, below the level of its tracks.

The determination of the excavability of rock tends to be more problematic than for soils and the cost of excavating rock may have a significant effect on the cost of the project as a whole. Rock is commonly defined contractually in a number of ways, the most common of which are by defining rock as a material which

(a) is present in specific geological horizons, such as Carboniferous Limestone, Dolerite, etc.,
and/or
(b) has a minimum strength, which is usually defined in terms of the unconfined compressive or point load strengths,
and/or
(c) is greater than a nominated minimum size, for example, boulders over $0.2m^3$,
and/or
(d) requires special plant or equipment for its removal, such as the use of wedges, rippers, blasting or approved pneumatic tools.

Thus, rock as defined contractually may or may not be equivalent to rock in a geological or engineering sense. It may therefore be necessary, in the planning, pricing or execution of earthworks, to assess the proportion of the rock, as defined contractually, that must be treated as rock in an engineering sense.

In engineering terms rock will require ripping, blasting or penumatic or hydraulic tools (or similar) for its removal, and the use of these techniques is more costly than those required for soils. The major factors controlling the excavatability of rock are the intact mass strength of the rock and the spacing, orientation and roughness of the discontinuities within the rock mass. Certain siliceous rocks such as sandstone and quartzite may be very abrasive to the cutting and tracking parts of the excavating plant, and this may be a significant cost item. The ability to excavate rock by the use of rippers decreases as the hardness of the rock and spacing of the discontinuities increase, and for a given rock in shallow excavations (up to 20 m deep) the ease of ripping decreases with depth as the rock becomes less weathered. The assessment of the rippability of rock, or whether blasting may be required, can be carried out by experienced engineers or engineering geologists from an assessment of the rock properties, or from charts relating

Table 3. Typical densities and bulking factors

Material		Bulk density (in situ) Mg/m^3	Bulk-up on excavation, %
Soils			
Sand, uniform —	loose	1.60 – 1.90	10 – 15
	compact	1.65 – 2.10	
Sand, well graded —	loose	1.75 – 2.20	10 – 15
	compact	1.90 – 2.25	
Gravel		1.70 – 2.25	10 – 15
Sandy gravel		1.90 – 2.25	
Clay —	soft	1.60 – 1.95	20 – 40
	firm	1.75 – 2.10	
	stiff	1.80 – 2.25	
	gravelly	1.65 – 2.30	
	organic	1.40 – 1.60	
	glacial	1.75 – 2.10	
Loam		1.50 – 1.60	25 – 35
Peat		1.05 – 1.40	25 – 45
Topsoil		1.35 – 1.40	25 – 45
Rocks			
Granite		2.60 – 2.70	50 – 80
Basalt/Dolerite		2.70 – 2.90	50 – 80
Gabbro		2.80 – 3.00	50 – 80
Gneiss		2.70 – 2.90	30 – 65
Schist and Slate		2.70 – 2.90	30 – 65
Quartite		2.60 – 2.75	40 – 70
Sandstone		2.45 – 2.65	40 – 70
Limestone		2.40 – 2.70	45 – 75
Marble		2.60 – 2.80	45 – 75
Chert and Flint		2.50 – 2.60	40 – 70
Marl		1.90 – 2.35	25 – 40
Shale		2.15 – 2.60	30 – 65
Chalk —	Upper and Middle	1.65 – 2.05	30 – 40
	Lower	2.00 – 2.40	
Coal		1.25 – 1.60	35

seismic velocities to rippability if such information is available; Fig. 1 is an example of such a chart.

Bulking and shrinkage

During excavation, soil and rock are loosened and sometimes broken down leading to a bulking-up. Thus, one cubic metre of material in the ground will occupy more than one cubic metre in the transporting vehicle, and this must be allowed for in assessing the amount of transport required and therefore the cost of transporting material. Table 3 shows typical values of in situ bulk density and bulk-up during excavation for soil and rocks present within the UK.

After the material has been laid down, and during compaction, the process is reversed and the material is rendered more dense. Depending on the nature of the materials and the degree of compaction there may be a net bulk-up or net shrinkage after the completion of compaction. This is not to be confused with wastage of fill as occurs in creating haul roads, overfilling embankments and the removal of material from site after contamination by rain, and which can be as great as 10 to 15% (although 5% is generally quoted as typical). The net bulk-up of most soils and weak rocks tends to be between 0 and 10%, and between 5 and 20% for harder rocks. Where water drains from the material, where the structure of the material is destroyed, or where a more dense arrangement of the soil particles is achieved, a net shrinkage may occur. Materials that commonly show a net shrinkage are chalk (0 to 15%) and sand (0 to 10%).

3. EARTHMOVING PRACTICE

The processes whereby soils or rocks are excavated, loaded into vehicles and transported to and deposited on embankments and tips are collectively called 'earthmoving'. Although the compaction of earthworks materials on the embankments or tips is often loosely considered to be part of the earthmoving processes, it is better regarded as a separate operation. If earthmoving is to be carried out efficiently it is important to determine the types of plant that can be used for the job in hand and to plan and organize the use of the plant so that the work can be carried out effectively and at the least cost.

Types of earthmoving plant

A wide variety of types of earthmoving plant is currently used in the UK by the construction industry. Some have been developed primarily for the industry, whilst others have been adapted from uses in other industries such as mining and quarrying. The choice of plant is further increased by the large number of international, multinational and national manufacturers of equipment. American, Japanese, German, Italian, Swedish and British designed and/or manufactured plant are commonly used in the UK and may be found working side by side. In practice, however, the major plant operators tend to use plant that is supplied by a limited number of companies in order that the equipment is standardized and the maintenance costs minimized.

Earthmoving plant can be classified in a number of ways. Here they are sub-divided on their primary roles in the

Fig. 2. Caterpillar D9H dozer with single shank ripper (Photo by courtesy of H. Leverton and Company Ltd)

earthmoving processes.

(a) Plant involved only in the excavation of materials.
(b) Plant that can only excavate and load materials.
(c) Plant that can only haul and deposit materials.
(d) Plant that can excavate, load, haul and deposit materials.

Table 4 summarizes the common types of plant used for earthmoving in the UK construction industry.

Plant that is involved in excavation only

Rippers. Rippers are fitted to dozers (see page 16) to loosen soil or weak rock so that it can be more easily excavated by scrapers, etc., or to loosen rock either in lieu of, or after, blasting (Fig. 2). In some cases rippers can be fitted to other types of plant for the same purposes but this is generally less effective. A ripper consists of one to three ripper shanks and teeth that are carried via brackets to

Table 4. Classification of earthmoving plant

Excavation	Excavation and load	Haul and deposit	Excavation, load haul and deposit
Rippers	Dragline	Dumpers	Dozers
Drill and blast	Face shovel	Dump trucks	Tractor drawn scrapers
Impact hammers	Forward loaders	Lorries	Motor scrapers
Hydraulic breakers	Grab	Conveyors	Dredgers
Graders	Back-hoe		
Skimmers	Bucket wheel excavators		

a cross bar that is lifted or lowered by hydraulic rams or cables. When rippers are used it is then necessary to reinforce the rear of heavy tractor units to take the strains induced during ripping. The shanks may be straight or curved and are capped with detachable teeth which protect the shank from wear and through which ripping force is generated.

Scarifiers and rooters. These are similar in concept to rippers and are designed for ripping paved surfaces and removing tree roots and obstructions, respectively.

Drill and blast. Where rock is encountered that cannot be economically excavated by rippers it may be necessary to use explosives to loosen and break up the rock.

(a) Drilling is carried out by using hand operated drills or by drifter or track-mounted rotary-percussion rigs using down-the-hole or top hammer drills to form the borehole (Fig. 3).

(b) Explosives usually consist of medium strength nitroglycerine gelatines such as Opencast Gelignite, ANFO (ammonium nitrate plus fuel oil) or slurry explosives such as Supergel. The blast is generally initiated by electric detonators or detonating fuse, and short delay detonators or detonating relays are commonly used. In excavations of limited size compressed gases (e.g. Cardox) or hydraulic splitters can be used in lieu of explosives to break out or loosen rock.

Impact hammers. These consist of a compressed air or diesel powered hammer that generates high frequency impacts to the ground via a steel tooth. They are commonly attached to

Fig. 3. Atlas Copco ROC 601 rotary percussion drill rig (Photo by courtesy of Atlas Copco (Great Britain) Ltd)

Fig. 4. Hymac 590C tracked back-acter (Photo by courtesy of Hymac Ltd)

the boom of crawler-mounted excavators and are used to break up limited quantities of rock, concrete etc., and are especially useful in confined areas.

Graders. Commonly used to maintain haul roads. They consist of a blade which can rotate in a circular arc about a horizontal axis and which is supported beneath a longitudinal frame joining the front steering and rear drive wheels. The front wheels are generally articulated whilst the rear wheels are set in tandem beneath the motor and control unit. The blade is used to trim and redistribute soil and the unit is therefore usually operated in the forward direction.

Plant that can excavate and load

The back hoe or back-acter. This consists of an articulated boom at the end of which is a bucket. The unit is generally set on a pneumatic-tyred wheeled, or a crawler, tractor unit. The boom, bucket arm and bucket are usually controlled by hydraulic rams although cables were commonly used in the past (Fig. 4). Back-acters operate by digging towards the machine in an arc from a small distance above the ground level to a position vertically below the outer edge of the machine. The maximum depth of excavation is related to the length of the boom and machines with depth capacities of between 2.5 m and 6 m are in common use. Loading is carried out by swinging the boom or turning the machine away from the working face to awaiting haulage vehicles.

The face shovel or loading shovel. This is extensively used in quarries and pits and on construction sites and is useful in excavating blasted rock in cuttings, etc. It is constructed in a similar manner to a back-acter except that the boom and

Fig. 5. Caterpillar 980C wheeled forward loader (Photo by courtesy of H. Leverton and Company Ltd)

bucket arm operates in the opposite direction, i.e. up and away from the machine. It is therefore used for excavating stockpiles or faces up to a height of about 10 m. Loading is carried out in a similar manner to the back-acter. Some of the larger excavators can be readily converted from back-acters to face shovels.

A forward loader. Consists of a pneumatic-tyred or crawler tractor at the front of which is mounted a wide bucket that can be moved in the vertical plane (Fig. 5). Excavation is carried out by driving the machine towards, and the bucket into, the material; the bucket is then turned and lifted upwards thus catching and excavating the material. In order to load the material to the hauling vehicle it is necessary to drive the machine towards, and empty the bucket over, the vehicle. They are generally used to excavate material at and for a distance above ground level and can be used to push or haul material over short distances in the loading bucket. Modern forward loaders have hydraulically powered loading buckets and many of the smaller units are equipped with a back-acter unit.

Draglines. Dragline equipment is operated from cranes or similar plant with a long boom. Excavation is carried out by pulling a bucket which is suspended on a cable towards the machine by a second cable. Thus the machine is suitable for excavating soft or loose materials at a level beneath or slightly above the level of its tracks. Excavated material can be removed to stockpiles or unloaded into haulage vehicles by rotating the machine with the bucket in the upright position to the required location, then upturning the bucket, (Fig.6).

A grab. Consists of a cable or hydraulically controlled bottom opening bucket that is suspended from the boom of a crane or similar plant. The bucket is opened and dropped onto the material to be removed; it is then closed and the material that is caught between the jaws is lifted up in the grab bucket and discharged into stockpiles or awaiting haulage vehicles. Grabs are therefore used for the excavation of pits or trenches, and loading to and from stockpiles.

Bucket wheel excavators. These are not commonly used on construction sites within the UK except for trenching work. They generally consist of a series of buckets set on a circular wheel or in a closed loop which is itself set on a boom that can move laterally and vertically. Material is excavated by the teeth of the buckets moving in an upward direction and discharged from the buckets on their downward course. The whole machine is usually supported by, and moves forward on, crawler tracks and therefore such machines tend to be used for the excavation of linear excavations such as canals and trenches.

Plant that hauls and deposits

Road lorries. Where haulage is required to take place on public roads it is necessary to use licenced and taxed lorries. These are available in varying sizes up to 32 tonnes capacity and generally have steel or steel/aluminium sheeted bodies. Such vehicles require to be loaded by other plant but can be unloaded by rear or side tipping, (Fig. 6).

Unlicenced lorries. These can be used when public roads are not traversed. They may often be old and may not be in a state of repair necessary to operate on public roads.

Fig. 6. NCK Rapier 406 crawler dragline discharging to tipper (Photo by courtesy of Ransomes and Rapier Ltd)

Fig. 7. B.M. Volvo DR 860 articulated dump truck (Photo by courtesy of Anglo Swedish Equipment Ltd)

Dump trucks and dumpers. Generally vary in size from 1 to 77 tonnes capacity, although larger capacity machines are manufactured and are generally used in large mines or quarries. Smaller dump trucks may be articulated and powered on all wheels and are suitable for the haulage and deposition of soft deposits (Fig. 7). Bodies of dump trucks may be heated by the exhaust from the engine and the sides are often splayed outwards to provide a large tipping target. The speed of tipping is increased over a road lorry by the absence of a tailgate. Small dumper units are available for work on small sites and commonly have the load carried in front of the driver.

Conveyors. Have been used with varying degrees of success on construction projects in the UK. They are built up with a number of units of endless flat belt conveyors placed in series and major changes in direction can be made at transfer points where material from one belt falls and is channelled onto the next. Loading is generally carried out via a hopper which may be designed to screen out over-sized material. The conveyor commonly ends in a stacker which allows the material to be spread over a wide area. Conveyors are generally used in quarries and in areas of very poor or problematical access or where the terrain is too steep for lorry access. The capital costs to set up a conveyor belt system are generally large in comparison to conventional plant, but operating costs are generally lower.

Plant that excavates, loads, hauls and deposits

Bulldozer (or dozer). A bulldozer is a tractor equipped with a front pusher blade which can be raised and lowered by hydraulics or cables (Fig. 2). An angle-dozer has a pusher blade which is capable of being set at an angle to push material sideways whilst the tractor moves forwards. The tractor unit is usually mounted on crawler tracks thus allowing it to travel over a wide variety of ground conditions, although wheel-mounted units are available. Blades are manufactured in a number of styles, but all require to be of heavy duty construction and have a hardened steel, basal leading edge, which is driven into the ground to cut and push the material to be excavated. Dozers have a variety of roles including excavating soils and weak rocks, ripping, moving excavated materials over short distances and acting as a pushing unit to boost the effective power of scrapers, etc. A wide range of crawler units are available ranging between 60 and 700 hp. Dozer blades may be attached to other plant (such as self-propelled compactors) so that they can operate without the need for a dozer.

Scrapers. These can excavate, load, haul and deposit material in one cycle and may be towed or self-propelled (motorized). They consist of a centrally mounted bowl, the bottom leading edge of which has a cutting edge, and the position of which can be controlled. Both towed and self-propelled scrapers are effectively articulated between the steering or towing portion and the bowl, and larger self-propelled scrapers may have a second engine mounted at the rear. During excavation and loading the tailgate is closed and the cutting edge lowered to the ground surface; the bowl is lowered and the cutting edge scrapes the top layer of soil into the bowl. As material builds up in the bowl it may be necessary to add further power by using a dozer behind acting as a pusher. On the completion of loading the apron is lowered and the bowl raised. Ideally haulage should take place over well maintained haul roads with the minimum of steep gradients or sharp turns. Deposition of the material is carried out by moving the tailgate forward with the apron raised and bowl partially depressed. The thickness of the layer of deposited soil can be regulated by the control of the tailgate, apron and bowl level.

Towed scrapers. Typically range in size from 5.4 m^3 to 16.8 m^3 struck capacity, and are generally towed by crawler tractors. They are limited to economic hauls of approximately 400 m each way.

Single engine motorized scrapers. Typically range in size from 10.7 m^3 to 24.5 m^3 struck capacity (15.3 m^3 to 33.6 m^3 heaped) (Fig. 8).

Double engine motorized scrapers are similar. Motor scrapers can generally be economically operated on hauls of up to 2.6 km each way, with optimum performance on hauls of approximately 800 m.

Elevating scrapers. Similar to conventional scrapers except that a rotating elevator within the bowl drags the material back from the front of the bowl, breaks it up and deposits within the bowl, thus reducing the resistance on loading. They generally range in size from 7.2 m^3 to 26 m^3 heaped capacity and may be self-propelled or towed units.

Dredgers. These are used for excavation below water and are usually purpose made floating vessels. Several different types are available including cutter-suction, bucket wheel, grab and dipper (face shovel) dredgers. Material can be pumped or transported by barge or the dredger to the area of deposition or off-loading point. The types and application of dredgers constitute a varied but somewhat specialized subject and it is not intended to consider them further in the context of earthmoving.

Planning of earthmoving

The planning of earthmoving should be carried out at the tender stage in order that the programme and estimate for the work can be carried out, and at the construction stage to determine the types of plant required and to assist in their management during the work.

The main objective in the planning of earthworks is to

Fig. 8. Caterpillar 631C single engined scraper (Photo by courtesy of H. Leverton and Company Ltd)

analyse the work to be carried out and determine the types and number of plant that would be most suited to the sequence of operations. From this the cost can be determined. *As no two contracts are the same, the analysis of the work and the production rates of the various types of plant that could be used may vary considerably and must be determined for each scheme.* The safest procedure is to work from first principles for each scheme, applying experience from past contracts wherever possible and for this it is necessary to determine the major factors that may control the work.

Factors controlling the planning and execution of earthmoving schemes

The major factors affecting the planning of earthmoving schemes and thus the choice of plant are

(a) material factors, such as the types, properties and variability of the materials to be excavated and transported
(b) geographical factors, such as the features and relief of the area, the means of access to and along the site, and the anticipated weather conditions during the working period
(c) spatial factors, such as the volumes, area and disposition of the materials to be excavated, and the time available for the various operations
(d) plant factors, such as the types of plant suitable and available to carry out the works, their operating characteristics, outputs and unit costs
(e) other factors such as legal, contractual or political restraints that may influence the selection, use or access of plant and materials.

It is therefore necessary to determine the nature, quantities and disposition of the materials to be excavated and the nature and distance of any haulage that may be involved, so that a choice of the types of suitable plant or plant teams can be made; from this, as assessment of the most economic plant teams can be made within the legal and contractural restraints.

Material factors

The nature of the materials present within a site affect the use and methods of excavation and transportation of the materials. If there is information available on the materials in site investigation reports, or reports from other sources, it is often advisable to make an assessment of the nature and distribution of topsoil, soil and rock (in both contractual and engineering senses) and further to identify whether each type of material will be suitable for use in the works, or is to be disposed of.

Topsoil is generally billed and treated separately from the other materials as it is required to be retained for landscaping the site after the completion of earthworks, or sold if it is not required.

Soils may be cohesive or granular and can generally be excavated without the use of rippers, blasting or pneumatic tools. Harder cohesive soils and coarser or denser granular soils may behave like weak rocks and their excavation may be assisted by ripping or the use of pneumatic tools,

Table 5. Rolling resistance of wheeled plant (by courtesy of Caterpillar Overseas S.A.)

The rolling resistance is the force that must be overcome to pull or roll a wheel over the ground.
Rolling resistance = rolling resistance factor x gross vehicle weight.

Haul road conditions	Rolling resistance factor	
	kg/t	As equivalent gradient
Hard smooth, stabilized surfaced roadway, no penetration under load, well maintained	20	2%
Firm smooth, rolling roadway with dirt or light surfacing, some flexing under load, periodically maintained	32.5	3%
With snow – packed	25	2.5%
loose	45	4.5%
Dirt roadway, rutted, flexing under load, little maintenance, 25 mm to 50 mm tyre penetration	50	5%
Rutted dirt roadway, soft under travel, no maintenance, 100 mm to 150 mm tyre penetration	75	7.5%
Loose sand/gravel	100	10%
Soft muddy rutted roadway no maintenance	100 – 200	10 – 20%

especially in excavations of limited size.

Rock in an engineering sense may not be the same as that defined in the specification for payment purposes. Weak rocks such as siltstone, mudstone and chalk, and those rocks containing numerous discontinuities (bedding planes, joints etc.) such as slate and sandstone, may be ripped. Harder rocks such as dolerite, granite etc., and poorly fractured rocks, such as some limestones and sandstones, commonly require blasting. The assessment of when blasting will be required is best based on experience although a good guide can be obtained from published charts if the seismic velocities of the various strata are known. An example of this is shown in Fig. 1. Rocks such as siliceous sandstone and quartzite may be highly abrasive and high drill bit and ripper tooth wear may be expected.

The density or unit weight of the materials to be excavated controls the amount of material that can be dug or transported, and the speed of loading. In addition the amount of bulk-up on excavation will affect the amount of haulage plant required. In Table 3 the densities and bulk-up for typical soils and rocks are presented.

The groundwater conditions within a site may have a significant effect on the types and production rates of the excavating plant, especially in granular soils such as sands and gravels. In the assessment of the groundwater conditions, it may be necessary to distinguish between perched or local water tables and general water tables and to predict the anticipated quantities of flow that may be induced. Where significant flows are likely to occur, some form of dewatering will be necessary, although in river and canal works, and in borrows or quarries, excavation below the water table may be possible by using draglines, back-acters or grabs.

The nature of soils or rock along the haul routes and areas of excavation should be studied as these can affect the type and operation of excavating and hauling plant. It may be important to determine whether the ground can support the plant that it is proposed to use (i.e. that it has an adequate bearing capacity) and to ascertain the likely tractive efficiency and rolling resistance of the plant, as shown in Table 5.

Topographic and environmental factors

The topography of the site may affect the manner in which excavation is carried out, the rate of production, the speed at which haulage vehicles can travel, and the accessibility of plant on part or all of the site. In addition it may also control the location of site offices, stockpiles and tips, the offices being best located on open and level ground, whilst tipping facilities may be more readily granted in valleys, depressions and old quarries.

The main environmental considerations include any limitations that may be placed on the plant employed by restrictions on noise and vibration levels or working hours and on the use of public roads. In some instances the type of vegetation covering the land may provide limitations to access until cleared, as is the case in wooded or boggy areas.

The weather conditions within a given site may also be an important factor. The balance between rainfall and evaporation is important in controlling the trafficability of the site. Rainfall tends to result in the softening of soils within the site and thus reduce the bearing capacity and trafficability and, in addition, may lead to excessively high moisture contents in the fill being compacted. This may lead to the cessation of work or to a reduction in the productivity. The susceptibility of any given plant to the effects of rainfall on different materials varies: clayey and fine grained granular soil and fine grained weak rocks are more affected by rainfall than coarser granular soils or harder rocks. Dump trucks are more susceptible than scrapers to reductions in performance due to adverse weather conditions. Frost may also inhibit productivity by hardening the ground in the excavations and precluding the possibility of compacting fill.

The amount of productive time available per annum also varies throughout the country with the minimum losses in productivity being in areas of low rainfall, high evaporation and minimum of frosty days such as occur in south east England, whilst high losses occur in the higher areas of northern England and Scotland where it is colder and wetter.

Spatial factors

Spatial factors have a significant influence on the numbers and types of plant to be used and the duration of the work, and consequently have a large effect on the cost of the work. The main items include the volumes of the various materials to be excavated and transported on or off site, the distances over which the haulage is to take place on or off site, and the duration or timing of the work.

On a small site where only limited types of plant can be used, the distance of transport may be a small factor in the total works, whereas the timing may be important and in some cases the double-handling of material may be required. Where larger amounts of material have to be excavated and transported over varying distances, the volumes of materials involved and the relationship between volume of material and the distance of haulage are often significant factors in the choice of plant. In such cases mass profile or mass haul diagrams, as shown in Fig. 9, may be used to assist in determining the best methods of excavating and hauling the materials.

Plant factors

In order to determine the types of plant that may be used it is necessary to analyse the work as described previously and ascertain the operations to be carried out. When the output rates of the various suitable and available combinations of plant or plant teams have been calculated, the number of teams and cost of the earthworks can be determined. The output rates can be calculated from a critical appraisal of the manufacturers published output rates, or more satisfactorily from an analysis of each operation based on first principles, aided if possible with data obtained from field use in similar circumstances. Typical plant combinations are detailed in the next section.

Other factors

A variety of contractual or legal constraints may be applied to the earthworks construction. These may include restrictions in the plant that can be used, or the hours during which it may be used. A typical example of the former is for the excavation of chalk and similar weak rocks where recent

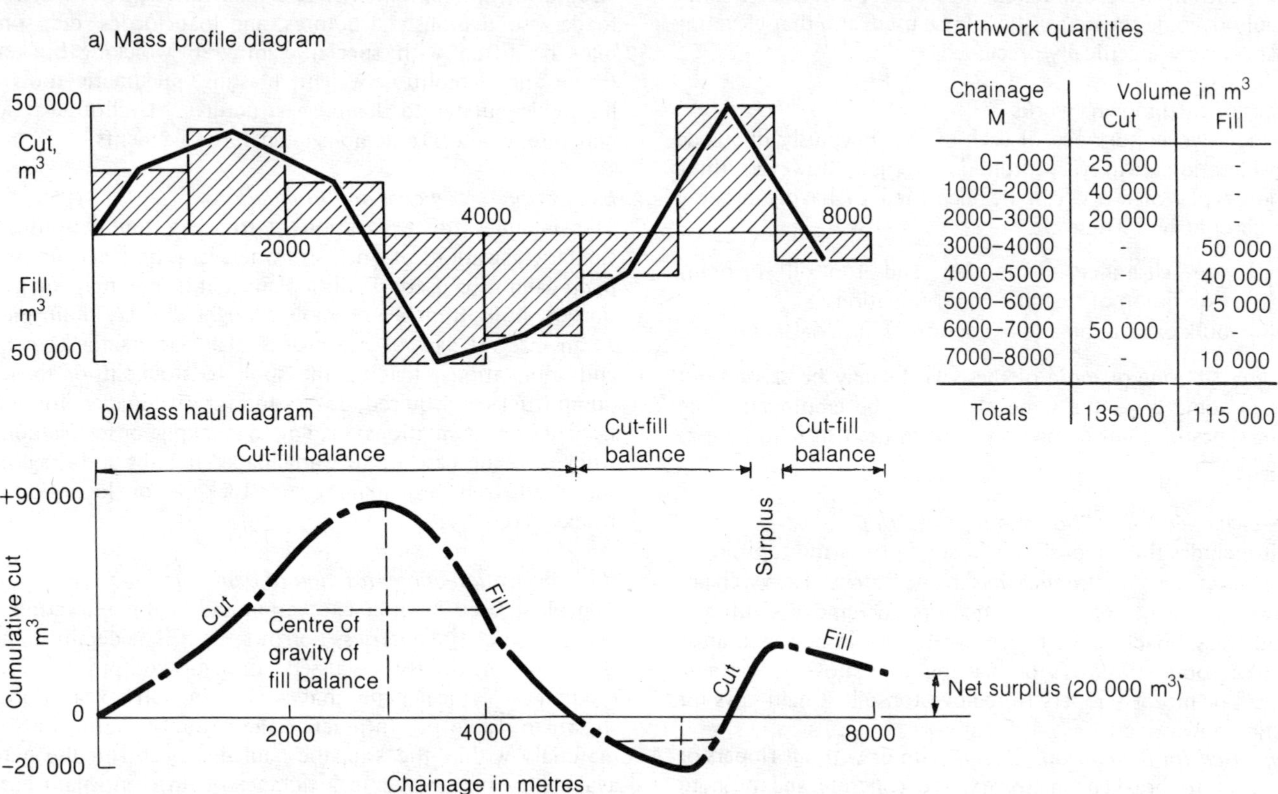

Chainage M	Volume in m^3 Cut	Fill
0–1000	25 000	-
1000–2000	40 000	-
2000–3000	20 000	-
3000–4000	-	50 000
4000–5000	-	40 000
5000–6000	-	15 000
6000–7000	50 000	-
7000–8000	-	10 000
Totals	135 000	115 000

Fig. 9. Mass profile and mass haul diagrams

specifications have stipulated that only plant that do not unduly degrade the material could be used, and that elevating scrapers were specifically precluded.

Types of earthmoving works

The relative importance of each of the previously discussed considerations will vary from job to job, however, three major types of work can be identified as having broadly similar controls. These are

(a) site clearance, demolition and top-soil stripping
(b) excavation of trenches and foundations
(c) bulk excavation and extraction of materials.

On any site one or more of these works may be carried out and in some cases the same plant may be employed. The main types of plant commonly used in each situation are as listed below.

Site clearance, demolition and topsoil stripping

This includes the removal of light scrub, trees and topsoil.

Plant used for site clearance and topsoil strip. Heavy chains drawn between two crawler tractors to remove scrub and small trees. Back-acters to remove topsoil in confined areas or tree roots. Rooters or blasting to remove tree roots. Towed or motor scrapers to remove topsoil. Chain saws for cutting down trees.

Plant used for demolition. Dozers with or without rippers or scarifiers to break up masonary and concrete and push to stockpiles. Back-acters which may also be fitted with attachments such as grabs, impact hammers, etc. to demolish structures, and pavements, and load lorries, etc. Forward loaders to demolish structures, and load lorries, etc; some may be fitted with special clam grab wrecking buckets. Crane and demolition weight, blasting, pneumatic tools or hydraulic burster to demolish structures. Guillotine-action concrete breakers to demolish concrete pavements.

Excavation of trenches and foundations

This includes the excavation of trenches, foundations and similar excavations of small size where large bulk-earthmoving plant cannot be used. Typical plant that is used may include: purpose-built trenches or mole ploughs that lay drain pipes as an integral part of the process; back-acters for trenches and foundations, loading the spoil to stockpile or lorries/dump trucks as required; dozers and forward loaders for wide trenches or foundations, loading to stockpile or lorries/dump trucks (in the case of forward loaders); grabs and draglines for deep trenches, loading to stockpile or lorries/dump trucks as required.

Bulk excavation and extraction of materials

The plant used in bulk excavation and in the extraction of soil and rock in quarries, borrows, etc., is generally larger than that previously discussed, as higher output rates are required. Typical plant that is used in bulk excavation is shown in Table 6. In practice the natural variations in the materials within the site, the haul distances and the plant availability may result in a number of different plant combinations excavating similar or different materials in the same general area.

Table 6. Plant typically used in bulk excavation

Excavation	Loading to	Uses
Bulldozer/Angledozer	–	Levelling, excavating and assisting other plant
Bulldozer/Angledozer	Stockpile	Pushing over short distances to stockpile
Bulldozer with ripper	–	Loosening and excavating rock or hard/dense soils
Drill and blast	–	Loosening and breaking of rock or hard/dense soils
Wheeled forward loader	Dump trucks, lorries, stockpiles, conveyors	Excavation of stockpiles, blasted rock, soil, faces, etc. where there is good running surface
Tracked forward loader	Dump trucks, lorries, stockpiles, conveyors	Excavation as above where the running surface is softer
Back-acter (wheeled or tracked)	Dump trucks, lorries, stockpiles	Excavation of faces of soil or weak rock, blasted rock, stockpiles, etc., usually sited above the material
Face shovel	Dump trucks, lorries, stockpiles	Excavation of faces of soil or weak rock, blasted rock, stockpiles, etc., usually sited below the material
Draglines	Dump trucks, lorries, stockpiles, conveyors	Excavation of soft soils to weak rocks especially where large reach is required or below the water table
Bucket wheeled excavator	Dump trucks, lorries, conveyors	Generally long term continuous linear excavation in soils or weak rocks
Towed scrapers	Self	Excavation of soil and weak rock in unconfined areas. Haul distances usually limited to 400 m
Motor scraper (single or twin-engined)	Self	Excavation of soil and weak rock in unconfined areas on material of subgrade strength of approximately 70 kN/m^2 or greater. Haul distances usually limited to 2.6 km
Elevating scraper	Self	As for motor scraper but not suitable for weak degradable rocks such as chalk
Dredgers	Self, barges, pumping	Excavation of soils and weak rocks below water

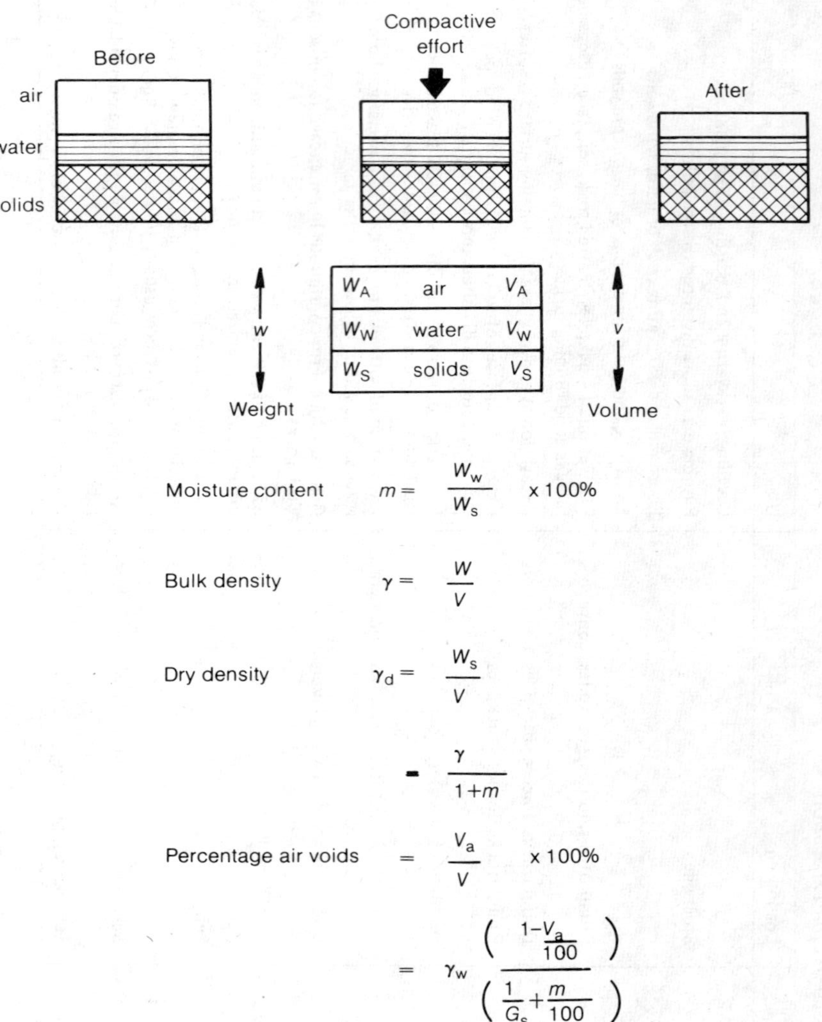

Fig. 10. Phase relationships in soil

The balancing of transporting plant to excavators, and pushers (dozers) to scrapers is often crucial to the efficient operation of bulk earthworks. Too many or too few dump trucks feeding off excavators, or dozers available to push scrapers, will leave plant standing idle and therefore not carrying out productive work. For example, one back-acter may be adequately serviced by two dump trucks where the haul distance is 100 m but five trucks may be required for a haul of 2 km.

4. COMPACTION PRACTICE

Theory of compaction

Soils can be considered to consist of three components or phases

(a) solid soil particles
(b) liquid, usually water, in the voids between the soil particles
(c) gas, usually air, in the remaining voids between the soil particles.

If no gas is present the soil is said to be saturated, and if it contains no liquid it is a dry soil.

Compaction is the process whereby the soil particles are mechanically made to pack more closely through a reduction in the air voids. The amount of compaction that can be achieved depends on the compactive effort applied and the pre-existing density, moisture content and percentage of air voids in the soil. The volume percentages of the three phases can be determined from the definitions given in Fig. 10 and a knowledge of the dry density and moisture content of the soil and specific gravity of the soil particles. Thus the results of compaction can be measured in terms of the dry density and moisture content, as for practical purposes it can be assumed that the specific gravity of a given soil does not change during the compaction process.

If soil is subjected to a given amount of compaction over a range of moisture contents a curve similar to that shown in Fig. 11 will be obtained, such that there is a unique value of the moisture content (the optimum moisture content) at which a maximum dry density can be achieved. In principle, when the moisture content is low the soil is stiff and difficult to compress and consequently a low density and high air voids are achieved. As the moisture content is increased, the water acts as a lubricant causing the soil to soften and become more workable, so that higher densities and lower air voids are achieved. As the air content decreases, the air and water tend to keep the soil particles apart thus precluding any further decrease in the air voids, however the total (air and water) voids increase with increasing moisture content and the dry density is reduced.

In 1933 R.R. Proctor introduced a compaction test in which the relationship of moisture content and dry density are determined in a laboratory under standard conditions. In the UK three compaction tests are described in BS 1377, *Methods of Testing Soils for Civil Engineering Purposes,* which are similar in principle but differ in the degree and

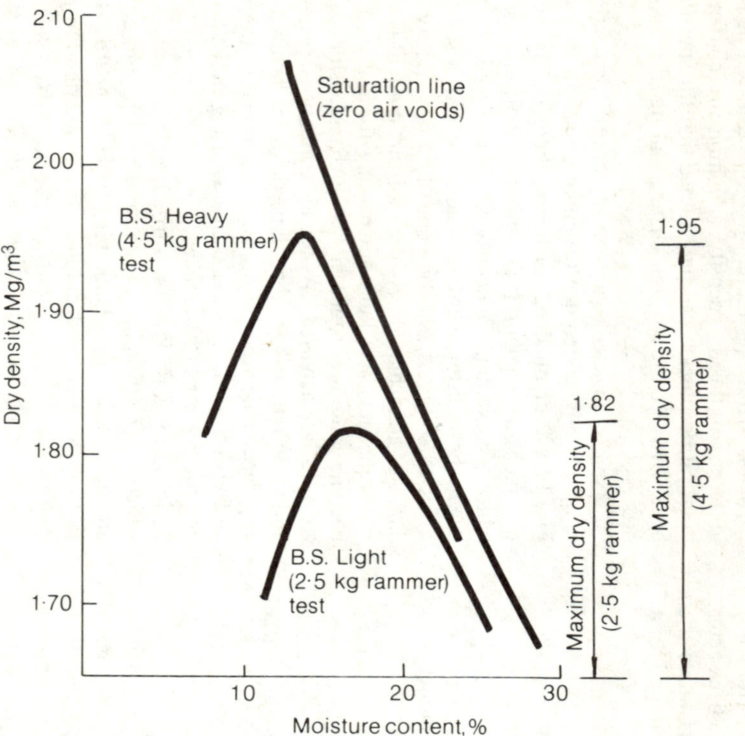

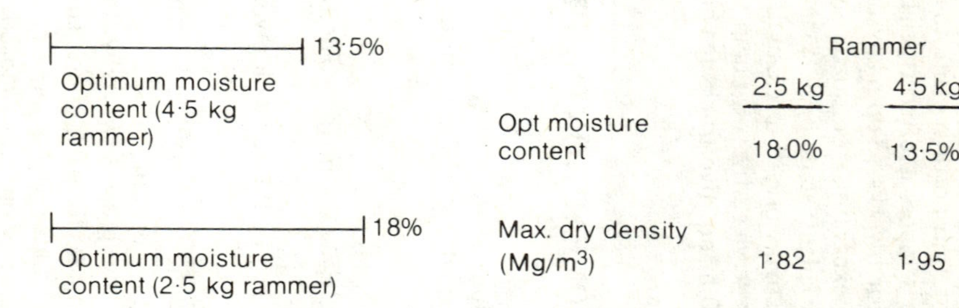

	Rammer 2·5 kg	Rammer 4·5 kg
Opt moisture content	18·0%	13·5%
Max. dry density (Mg/m^3)	1·82	1·95

Fig. 11. Relationships between moisture content and dry density for compaction test using 2.5kg and 4.5kg rammers

mode of application of the compactive effort. The British Standard (BS) (Light, Standard, Proctor or AAHSO) test involves compacting soil with a 2.5 kg rammer (as described in test No. 12), the BS Heavy (modified Proctor or modified AAHSO) test involves a larger sample of compacting soil with a 4.5 kg rammer (as described in test No. 13), whilst the Vibrating Hammer test for granular soils uses vibrational energy of known frequency applied for a fixed time period (as described in test No. 14).

In general, cohesive soils tend to have a lower maximum dry density and higher optimum moisture content than granular soils as is illustrated in the typical values given in Table 7. In addition the greater compactive effort of the BS Heavy compaction provides a higher maximum dry density and lower optimum moisture content than the BS test, as is illustrated in the example given in Fig. 11.

Reasons for compaction

Compaction is one of the most important aspects in the construction of structural embankments, such as road embankments or earth dams. The process of compaction has three beneficial effects on the soil, it reduces its compressibility and permeability and increases its strength. Uncompacted soil is compressible and will settle under loading and this provides a greater tendency for changes in the moisture content of the fill. This can in turn lead to excessive settlements under loading and loss of strength of the material, which can be minimized with adequate well controlled compaction.

Control of compaction

The degree of compaction obtained in the field can be monitored by determining the in situ (or field) density and moisture content and comparing them to the maximum dry density and optimum moisture content as determined from standard laboratory tests on samples of the material in question.

There are five commonly used methods of determining the in situ density.

(a) Core Cutter Method [BS 1377 (1975) Test 15D].
(b) Sand Replacement Method [BS 1377 (1975) Tests 15A to C].
(c) Immersion in/Displacement of Water Method [BS 1377 (1975) Tests 15E and F].
(d) Rubber Balloon Method.
(e) Nuclear Methods.

Of these the first two methods are more commonly used in the UK. The moisture content and specific gravity (if required) can be determined from laboratory tests on material obtained during the density test.

On large earthworks sites there are often many soil types present, each of which will have its own compaction characteristics and it may therefore be necessary to determine the laboratory compaction characteristics and in situ densities of each type of material present.

The standard laboratory compaction tests require the size of granular material to be less than either 20 mm or 37.5 mm. For soils containing coarse granular material it

Table 7. Typical compaction test results

Material	Type of compaction test	Optimum moisture content, %	Maximum dry density, Mg/m³
Heavy clay	B.S. (2.5 kg)	26	1.47
	B.S. Heavy (4.5 kg)	18	1.87
Silty clay	B.S.	21	1.57
	B.S. Heavy	12	1.94
Sandy clay	B.S.	13	1.87
	B.S. Heavy	11	2.05
Silty gravelly clay	B.S.	17	1.74
	B.S. Heavy	11	1.92
Uniform sand	B.S.	17	1.69
	B.S. Heavy	12	1.84
Gravelly sand/sandy gravel	B.S.	8	2.06
	B.S. Heavy	8	2.15
	B.S. Vibrating hammer	6	2.25
Clayey sandy gravel	B.S.	11	1.90
	B.S. Vibrating hammer	9	2.00
Pulverized fuel ash	B.S.	25	1.28
Chalk	B.S.	20	1.56
Slag	B.S.	6	2.14
Burnt shale	B.S.	17	1.70
	B.S. Heavy	14	1.79

may be necessary to adjust the test results so that they compare more accurately with the field densities. Alternatively non-standard tests can be carried out using larger moulds and the compactive effort adjusted to the size of the mould.

Compaction plant

Although considerable compactive effort is available from the passage of earthmoving equipment, it is generally necessary to use one or more types of purpose-made compaction equipment to achieve the adequate compaction of soil and rock fills. Compaction plant can be divided into three main groups

(a) rollers – including smooth-wheeled, pneumatic-tyred and tamping rollers and construction traffic
(b) vibrators – including rollers and plates
(c) rammers – including power rammers, tampers and falling weights.

Rollers

These apply pressure to the ground whilst traversing forwards or backwards.

Smooth-wheeled rollers. These are widely used and consist of smooth-wheeled drum rollers that are manufactured as three-roll or tandem-roll self-propelled units, or as a single towed roller. They range in size from 1.7 to 17 tonnes deadweight, which may be increased by adding sand or water ballast to the rolls or by placing ballast weights on the frame of the roller. The load applied to the ground is related to the load per unit width of the rolls and it is therefore often necessary to determine this in addition to the gross weight of the roller. Where a roller has more than one axle the higher load per unit width is normally taken. Optimum speeds are generally from 2.5 to 5 km/h.

Pneumatic-tyred rollers. May be towed or self-propelled and may have one or two axles, on which are set a number of rubber-tyred wheels (Fig. 12). On two axle machines, one axle generally has one less wheel than the other and the tyres are set so that the ground is completely covered by the passage of the machine. The weight of the roller can generally be increased by applying ballast to a box or platform over or between the axles. The compactive effort can be further

Fig. 12. Hyster C530A pneumatic-tyred compactor (Photo by courtesy of Hyster Europe Ltd)

modified by adjusting the inflation pressures of the tyres. Provision is usually made for a certain amount of vertical movement of individual or pairs of wheels, to enable the roller to negotiate uneven ground whilst exerting steady pressure on the soil. Optimum speeds are generally between 1.6 and 24 km/h. The wobbley-wheel roller is a type of pneumatic-tyred roller in which the wheels are set to wobble from side to side, providing more kneading action to the soil.

Grid rollers. Have a cylindrical steel mesh roller and may be balasted with concrete blocks (Fig. 13). They are generally towed units and can operate at speeds between 5 and 24 km/h. Typical weights vary between 5.5 tonnes net to 15 tonnes ballasted.

Tamping rollers. Include sheepsfoot and pad rollers and consist of steel drums fitted with projecting feet (Fig. 14). They may be self-propelled or towed with single or multiple axles and may be ballasted as required. The shape of the projecting feet vary considerably and may be circular, square or rectangular and are available in different lengths and tapers. Optimum speeds vary between 4 and 10 km/h.

Construction traffic. Is similar in effectiveness to pneumatic-tyred rollers. However, in practice it is difficult to ensure complete and uniform coverage of the fill and this requires the plant to operate at less than normal speeds. Compaction achieved by construction traffic is therefore usually considered a bonus; it can however lead to extensive overcompaction, rutting and degredation of the fill if traffic is concentrated over limited areas.

Fig. 13. Hyster Model D towed grid roller (Photo by courtesy of Hyster Europe Ltd)

Fig. 14. Bomag K300 self-propelled tamping roller (Photo by courtesy of Bomag (Great Britain) Ltd)

Vibrating compactors
These apply both pressure and vibrational energy to the ground and this can be applied by either vibrating rollers or plates.

Vibrating rollers. Generally range in size from 0.5 to 17 tonnes (static) and may be towed, self-propelled (Fig. 15) or manually guided (Fig. 16). The former generally operate at speeds of between 1.5 and 2.5 km/h, whereas manually guided machines typically operate at between 0.5 and 1.0 km/h. Repeated passes at high speed are generally not as effective as fewer passes at lower speeds. Where tandem rollers are used, both can be considered to be applying compaction; the input of compacted effort is generally adjusted to that of the roll with the lesser weight per unit width. The frequency of vibration of a given machine is generally fixed by the manufacturer and typically varies from 20 to 35 Hz for larger rollers and from 45 to 75 Hz for smaller rollers, although rollers with variable frequencies have been developed. When the vibrating mechanism is not used the roller can be considered as a deadweight roller. The drums of vibrating rollers are usually of a smooth type, although vibrating tamping rollers are also manufactured.

Vibrating plate compactors. Vary in weight from 100 kg to 2 tonnes with plate areas of between 0.16 m^2 and 1.6 m^2. The smaller versions are manually guided and therefore suitable for compacting small or awkwardly shaped areas

Fig. 15. Bomag BW 212S self-propelled vibrating roller (Photo by courtesy of Bomag (Great Britain) Ltd)

Fig. 16. Bomag BW 90S manually guided vibrating roller (Photo by courtesy of Bomag (Great Britain) Ltd)

Fig. 17. Bomag P2000 vibrating plate compactor (Photo by courtesy of Bomag (Great Britain) Ltd)

(Fig. 17). They usually travel at about 0.7 km/h.

Vibrotampers. Usually weigh between 50 kg and 100 kg. Compaction is induced by vibrations that are set up in a base plate through a spring activated by an engine-driven reciprocating mechanism. They are generally manually guided and are not suited to heavy compaction, although they are suited for use in confined spaces.

Compaction by impact

Three types of plant are available for the compaction of soils and rock fill.

Power rammers. These are manually driven machines suitable for the compaction of small or awkwardly shaped areas. Explosions in an internal combustion engine cause the machine to be driven upwards and the subsequent impact of the base plate on the ground results in compaction. They generally weigh about 100 kg although special larger versions are available.

Weight dropping rammers. Consist of a weight of 180 kg or more that is set between guides and which can be dropped from variable heights of up to 3 m or so.

Dynamic consolidation. This involves the controlled dropping of weights from typical heights of between 15 m and 40 m, thereby subjecting the ground to impacts in the order of 150 to 500 tonne metres. The technique involves the use of large cranes and weights and involves the monitoring of water levels, densities, strengths and bearing capacity of the ground during and after compaction. It is a process carried out by a limited number of specialist contractors and is suitable for the compaction of larger areas of compressible fill of considerable depth (up to 30 m or so).

Specification for compaction

The two basic types of specification in use in the UK are Performance and Method specifications.

Performance specifications

When compaction is carried out to a performance specification, the contractor is required within limits to choose his own method of compaction, but must achieve the specified result. This is usually measured in terms of the density or air voids, or in some cases a bearing strength, as determined by, for example, plate loading tests. Densities of 90 to 95% of the maximum dry density achieved in the BS Heavy compaction test, or between 5% and 10% air voids, would be typical acceptable levels of compaction for soils. It therefore provides an effective means of producing an embankment of consistent and known properties.

Such a specification places a greater responsibility on the contractor to select the soils to be compacted, the appropriate compaction plant, thicknesses of layer to be compacted and number of passes of the plant. This may involve numerous trials to determine the optimum combinations for each soil type selected. In addition both the contractor and the supervising engineer will need to carry out regular tests to determine the degree of compaction and will therefore probably require on-site laboratory facilities. As a result such specifications tend to be adopted for the larger earthworks projects such as in dam or road construction.

Method specifications

Method specifications are extensively used in the UK for the control of compaction and are based on the Department of Transport *Specification for Roads and Bridges*, which may be adapted to meet the local conditions. This type of specification relates the type of plant, thickness of layer of soil or rock, and number of passes of the plant for each type of material, together with any other controls that may be required. Provided that the contractor has carried out the work according to the specification, any deficiencies in the work are the responsibility of the engineer, and it is he who has the greater need to monitor the methods used in, and results of, the work. The compaction requirements as set down in the Department of Transport *Specification for Roads and Bridges* (1976) are reproduced in Table 8.

Compaction practice

The selection of the appropriate plant to carry out compaction on a site can be important if the work is to be carried out economically and the required degree of compaction achieved. Four main factors can be identified for consideration.

(a) Material factors, such as the types, properties and variability of the material to be compacted.
(b) Spatial factors, such as the volume, area and disposition of the materials to be compacted and the time available for compaction.
(c) Plant factors, such as the types of plant involved or available, their operating characteristics and outputs, and the unit cost of plant.
(d) Factors such as legal or contractual limitations that may influence the selection or use of compaction plant.

Table 8. Department of Transport compaction requirements (By permission of the Controller of Her Majesty's Stationery Office)

D = Maximum depth of compacted layer, (mm) N = Minimum number of passes *Roller must be towed by track-laying tractor

Category	Cohesive soils		Well graded granular and dry cohesive soils		Uniformly graded material	
	D	N	D	N	D	N
SMOOTH WHEELED ROLLER						
Mass per metre width of roll						
2100kg to 2700kg	125	8	125	10	125	10*
2700kg to 5400kg	125	6	125	8	125	8*
over 5400kg	150	4	150	8	unsuitable	
GRID ROLLER						
Mass per metre width of roll						
2700kg to 5400kg	150	10	unsuitable		150	10
5400kg to 8000kg	150	8	125	12	unsuitable	
over 8000kg	150	4	150	12	unsuitable	
TAMPING ROLLER						
Mass per metre width of roll						
over 4000kg	225	4	150	12	250	4
PNEUMATIC-TYRED ROLLER						
Mass per wheel						
1000kg to 1500kg	125	6	unsuitable		150	10*
1500kg to 2000kg	150	5	unsuitable		unsuitable	
2000kg to 2500kg	175	4	125	12	unsuitable	
2500kg to 4000kg	225	4	125	10	unsuitable	
4000kg to 6000kg	300	4	125	10	unsuitable	
6000kg to 8000kg	350	4	150	8	unsuitable	
8000kg to 12 000kg	400	4	150	8	unsuitable	
over 12 000kg	450	4	175	6	unsuitable	
VIBRATING ROLLER						
Mass per metre width of a vibrating roll						
270kg to 450kg	unsuitable		75	16	150	16
450kg to 700kg	unsuitable		75	12	150	12

(Cont. on facing page)

Category	Cohesive soils		Well graded granular and dry cohesive soils		Uniformly graded material	
	D	N	D	N	D	N
VIBRATING ROLLER						
700kg to 1300kg	100	12	125	12	150	6
1300kg to 1800kg	125	8	150	8	200	10*
1800kg to 2300kg	150	4	150	4	225	12*
2300kg to 2900kg	175	4	175	4	250	10*
2900kg to 3600kg	200	4	200	4	275	8*
3600kg to 4300kg	225	4	225	4	300	8*
4300kg to 5000kg	250	4	250	4	300	6*
over 5000kg	275	4	275	4	300	4*
VIBRATING PLATE COMPACTOR						
Mass per unit area of base plate						
880kg to 1100kg	unsuitable		unsuitable		75	6
1100kg to 1200kg	unsuitable		75	10	100	6
1200kg to 1400kg	unsuitable		75	6	150	6
1400kg to 1800kg	100	6	125	6	150	4
1800kg to 2100kg	150	6	150	5	200	4
over 2100kg	200	6	200	5	250	4
VIBRO TAMPER						
Mass						
50kg to 65kg	100	3	100	3	150	3
65kg to 75kg	125	3	125	3	200	3
over 75kg	200	3	150	3	225	3
POWER RAMMER						
Mass						
100kg to 500kg	150	4	150	4	unsuitable	
over 500kg	275	8	275	12	unsuitable	
DROPPING-WEIGHT COMPACTOR						
Mass of rammer over 500kg						
Height of drop:						
over 1m to 2m	600	4	600	8	450	8
over 2m	600	2	600	4	unsuitable	

Thus it is usually necessary to determine the nature of the materials present so that the choice of the types of suitable compaction plant can be made; from this, an assessment of the most economic combination of plant types and sizes to be used within the legal or contractual constraints can be made. A simplified example is included in section 5.

Material factors

The most important material factors that influence the compaction of soils and rock are the types of material to be compacted and their moisture contents, strengths and gradings. In addition the degree of variability of the types and properties may also be important factors, especially when more than one type of material is present.

The type and to some degree the size of compaction plant can be broadly related to the materials to be compacted and, as a guide, a summary is included in Table 9. The DOT Specification (1976) classifies suitable materials broadly into 'Cohesive', 'Well graded granular and dry cohesive' and 'Uniformly graded' soils and 'Rock fill', for compaction purposes. In Table 6/2 of the Specification (reproduced in Table 8), the type of material and the thickness of the compacted layer are related to the number of passes required for different types and sizes of compaction plant. In addition, the contractor is afforded the right to carry out site trials to establish whether the same end result can be achieved using non-specified combinations of plant and compactive effort.

Where a range of materials with differing properties are present within a site, it may be found that several types of compaction plant would ideally be required. On large projects it may be economically viable or necessary to employ a number of different types of plant, but on smaller sites only one or two items of plant would generally be used. In practice a compromise is often reached whereby a limited range of plant is used with varying degrees of efficiency, and thus the unproductive standing or down time of the plant is minimized.

In recent years the developments in the design and manufacture of vibrating rollers have led to their adoption over wide ranges of materials, largely at the expense of deadweight rollers. Smooth-wheeled deadweight or vibrating rollers may be used in conjunction with other compaction plant, for example, to seal the surface at the end of the working period after tamping rollers have been used to prevent the collection of surface water in the imprints.

The compaction of weak rocks such as chalk, marl, mudstones and shales has often been problematical especially when they are weathered. Instability of the fill has frequently been caused by the breakdown of the structure of the material during excavation, haulage and compaction and the development of high pore pressures during compaction. The problems are more severe where the material is weathered, and where the natural blocky structure has been partially or totally naturally broken down. The potential for instability can be reduced by carrying out the work with the minimum of disturbance to the structure of the material (for example, by excavating it with back-acters or face shovels, transporting it in dump trucks and compacting it with grid or deadweight rollers, whilst ensuring the even

Table 9. Summary of general suitability of compaction plant for different material types

Compaction plant	Principle soil types							
	Cohesive		Granular					
	Wet	Others	Well graded Coarse	Well graded Fine	Uniform Coarse	Uniform Fine	Soft rock	Hard rock
Smooth wheeled roller		X	X	X			X	
Pneumatic-tyred roller	X		X	X	X	O	O	O
Tamping roller	X	X	O	X	O		O	
Grid roller		X	X	X	X	O	X	O
Vibrating roller	O	X	X	X	X	X	O	X
Vibrating plate		O	X	X	X	X	O	X
Vibro-tamper		X	X	X	X	X	O	X
Power rammer	O	X	X	X			O	O
Dropping weight		X	X	X			X	X
Dynamic consolidation	O	X	X	X	X	X	X	X

X = Most suited. O = Can be used but less efficiently.

distribution of construction traffic over the full width of the fill). In recent years significant progress has been made in developing techniques to predict the likely behaviour of weak rocks during earthmoving and compaction.

Spatial factors

In general the larger types of compaction plant have higher output rates than small plant as they are able to compact a thicker layer with fewer passes and have wider roll widths. However, they are less manoeuvrable and less able to work in confined areas.

It is therefore frequently necessary to make an assessment of the volumes of each type of material to be compacted and the time during which the compaction plant is required. In addition, an appraisal should be carried out of the topography and shape of the areas in which compaction is required, so that any limitations on the size of suitable plant can be determined. Large areas of fill would usually be compacted using the larger towed or self-propelled rollers (smooth-wheeled, grid, tamping, pneumatic-tyred or vibratory) whereas in smaller areas, such as backfilling around structures, drainage works, etc., small rollers, vibrating plates or tampers and power rammers are commonly used.

Plant factors

The previous appraisal of the material and spatial factors should have determined general guidelines by which the choice of the types and sizes of plant can be made. The final choice of plant or combination of plant that can be used most efficiently will be largely determined by the plant factors.

The output rates of compacted material can be obtained either from a critical appraisal of the manufactures rating of the plant, from a separate assessment of the time taken to carry out one pass of the plant over a typical distance, or from proven output rates obtained during the use of such plant in previous work.

The cost of carrying out the compaction with different types of plant can then be determined. The time required to carry out compaction with one unit of plant (t) can be determined for different locations and types of material from

$$t = \frac{\text{Volume of the material to be compacted}}{\text{Output rate of the plant for that material}}$$

The number of units (N) required to carry out the work in the time available (T) is

$$N = \frac{\text{Time required for one unit}}{\text{Time available}} = \frac{t}{T}$$

Hence the total cost of employing a certain type of plant is

Total cost (C) = number of units required x unit cost of plant (c) x time available

$$C = N \times c \times T = c \times t$$

This appraisal may require further modification to allow for the non-availability of the chosen plant at the time required, or the use of various types of plant within the same site. Furthermore, in those cases where small volumes are to be compacted or where compaction is required to be carried out intermittently, it may be necessary in the long term to use plant at less than peak efficiency, especially where that plant can be used for other types of compaction (such as compaction of sub-bases, etc.) during the period when it is not required on the earthworks. These considerations have been significant factors in the increasing popularity of vibrating rollers, as they can be used over a wide range of materials and therefore over a wide range of applications. For example, on small sites the same roller may be used for the compaction of bulk fill, hardcore and sub-base materials.

Contractual factors

In certain circumstances the choice of compaction plant may be limited by the requirements of the specification for the works.

An example of this is included in the Department of Transport Specification Clauses 6Q7.5 and 608.6, which require rock fill to be placed in layers not exceeding 450 mm (loose) depth, spread and levelled by a crawler tractor of at least 15 tonnes weight and compacted by at least 12 passes of a vibrating roller of at least 1800 kg/m width or a grid roller of at least 800 kg/m width. When rock is placed as general fill, these restrictions are not imposed and the rock must be compacted as a well-graded granular soil.

5. EARTHWORKS PROJECTS : EXAMPLES OF PLANNING & CONSTRUCTION

The factors that must be considered in the planning and execution of earthmoving projects and of the compaction of fill have been described in sections 3 and 4. In this section, simplified examples of the planning of earthmoving and compaction projects are presented, together with brief reviews of contracts that have been carried out.

Earthmoving

Example of the planning of an earthmoving scheme

A site 150 m wide and 250 m long is to be developed for warehousing units. This involves the stripping of topsoil, reducing the site level and removing the material off site. The following materials are to be excavated and disposed of

Clay	18 125 m^3
Topsoil	10 000 m^3
Total	28 125 m^3

The site is reasonably flat and has good access. Tipping space is available 3.0 km distant. The work is to be carried out in the early summer and approximately 10 weeks are available on the provisional programme; production is limited to a 40 hour week.

During excavation, bulking of the soils will occur and thus must be allowed for, thus

Loose volume = bulking factor x bank volume; therefore

Clay loose volume = 18 125 x 1.3 = 23 562m^3
Topsoil loose volume = 10 000 x 1.45 = 14 500 m^3

The estimator has chosen to investigate excavation using tracked forward loaders with 14 m^3 (20 t) road lorries for the removal of the soil off site.

For a 1.53 m^3 capacity loader

Bucket fill factor = 0.8 (80% capacity)
Load per cycle = 0.8 x 1.53 m^3 = 1.22 m^3
No. of loads per 14 m^3 lorry = 14 ÷ 1.22 = 11.5 (approx. 12)
Loading time of loader = 0.8 min/load (from manufacturer's tables)
Total load time/lorry = 12 x 0.8 min = 9.6 min
add time for lorry to position = 2.0 min
∴ Total load time = 11.6 min

Output per loader per 40 hour week working 50 min/h

$$= \frac{40 \times 50}{11.6} \text{ loads/week} \qquad = 172 \text{ loads/week}$$

$$= 2408 \text{ m}^3\text{/week}$$

Total time required for one loader = 38 062 ÷ 2408 weeks = 15.8 weeks

In order that the work can be completed within the allocation time period, two loaders will be required for 8 weeks each (thus allowing time for the establishment of the plant on site, etc.).

The lorries have the following working cycle

Positioning at the loader	2.0 min
Loading	9.6 min
Moving on/off site	1.0 min
Travel 2.5 km each way at 30 km/h	10.0 min
Tipping time	3.0 min
Total	25.6 min

Number of lorries required per loader = 25.6 ÷ 11.6 = 2.2

Total number of lorries required = 4.4

Allowing for breakdowns, etc. 5 lorries will be required; thus the total plant required will be 2 loaders and 5 lorries.

Example of an earthmoving project

An example of a large earthmoving project can be well illustrated by a motorway that has been recently constructed in South Wales, in which both soils and rock had to be excavated. Because of the scale of the project the example has been simplified without altering the fundamental pattern of the work.

The contract involved the excavation of three major and one minor cutting, the construction of five embankments and the disposal of unsuitable and surplus suitable materials in off-site tips, as is shown in Fig. 18. In the middle of the works lay a large viaduct which effectively limited the transport of fill along the line of motorway.

The mass haul indicated that short to medium hauls of up to 3km and long hauls of up to 10km would be required and that a mixed fleet of motor scrapers, dump trucks and lorries would be required. Rock was known to be present within two of the cuttings and would require dump trucks or lorries for its transportation. This requirement could be only partly accommodated by the requirements of the mass haul and consequently tended to reduce the amount of work that could be carried out by motor scrapers. In the event more rock was encountered than anticipated at the east end of the contract and further reduced the amount of work carried out by motor scrapers. In Fig. 18 the more important items of plant used to excavate, load and transport the rock and soil are shown together with the major hauls.

Compaction

Example of the planning of a compaction scheme

A series of embankments are to be constructed of rockfill and well graded granular soils. After the initial build up of plant, it is anticipated that between 15 000 m³ and 22 000 m³ of material is to be compacted per week. The types of rollers that should be used have to be determined. Rock fill and well graded granular soils can be treated similarly for compaction purposes and are generally best compacted by vibrating rollers. Therefore a study into whether a towed or self-propelled roller would be more suited is to be carried out.

For both types of roller the specification defines the number of passes for a given layer thickness of fill at an operational speed of 1.5 to 2.5 km/h (average 2.0 km/h), and an overlap of 0.15 m each side of the roll is to be taken on each pass. The sizes are

	Towed roller	Self-propelled double drum roller
Roll width	2.08 m	2.03 m
Weight/m width of roller	5600 kg/m	3000 kg/m

From the specification on compaction (D.O.T.) the following can be determined

	Towed roller	S.P. double drum roller
Max. thickness of compacted soil	275mm	200 mm
Min. number of passes	4	2 *

* For double drum rollers of equal mass/metre width the number of passes stated can be divided by two.

For towed rollers:

Area covered/h = effective width of roller x travel speed

$$= (2.08 - 0.3) \times 2.0 \times 1000 \text{ m}^2/\text{h}$$
$$= 3560 \text{ m}^2/\text{h}$$

$$\text{Total production/h} = \frac{3560 \times 0.275 \text{ m}^3/\text{h}}{4} = 244.75 \text{ m}^3/\text{h}$$

$$\text{Total production/week} = 244.75 \times 54 = 13\,217 \text{ m}^3/\text{week}$$

This makes no allowance for working at less than peak efficiency, starting, turning, breakdowns, etc. A correction is required to allow for 60% working efficiency.

$$\text{Corrected weekly production} = 13\,217 \times 0.6 \text{ m}^3/\text{week}$$
$$= 7930 \text{ m}^3/\text{week}$$

For self-propelled rollers:

$$\text{Area covered/h} = (2.03 - 0.3) \times 2.0 \times 1000 \text{ m/h}$$
$$= 3460 \text{ m}^2/\text{h}$$

$$\text{Total production/h} = \frac{3460 \times 0.2 \text{ m}^3/\text{h}}{2} = 18\,684 \text{ m}^3/\text{week}$$

Allowing for the efficiency correction as before

$$\text{Corrected weekly production} = 18\,684 \times 0.6 \text{ m}^3/\text{week}$$
$$= 11\,210 \text{ m}^3/\text{week}$$

To achieve peak production either 2 self-propelled tandem rollers or 3 towed vibrating rollers would be required. In this case the unit cost of operating a towed roller (including the crawler tractor) was 1.4 times that of the tandem roller and a significant saving could be made if two self-propelled tandem rollers were used. However, if the work required had

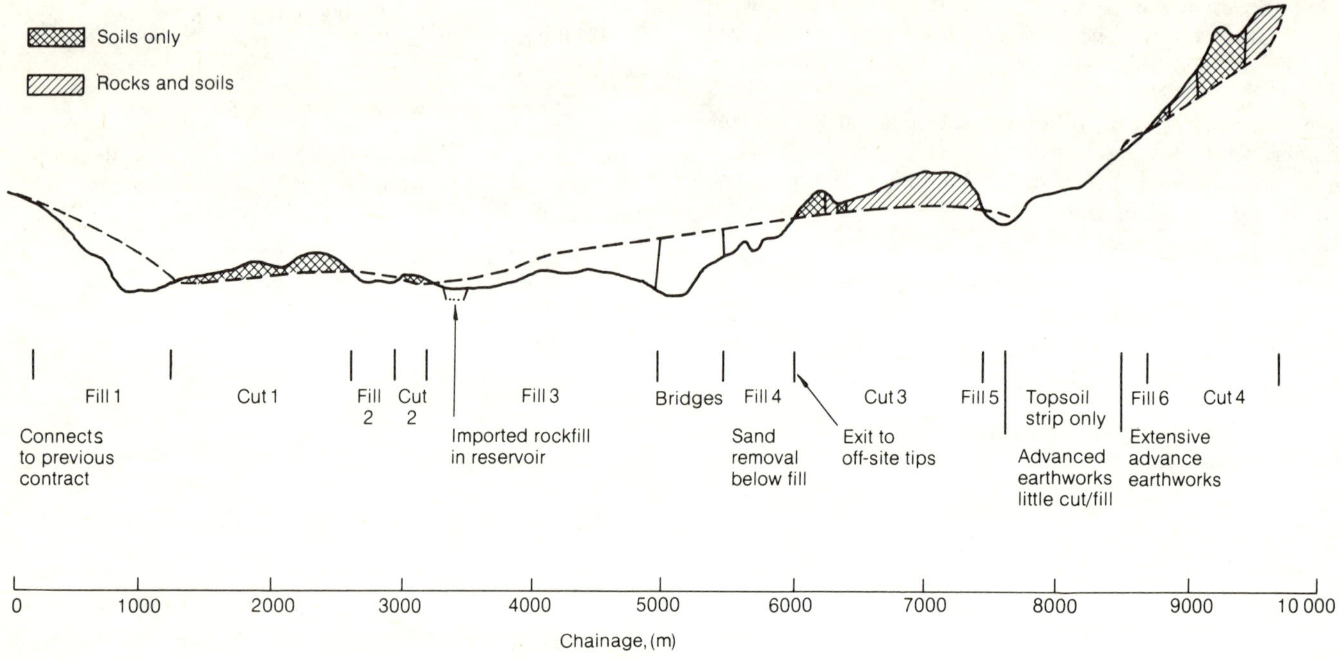

Fig. 18. Example of a large earthmoving project (cont. on facing page).

Item	Fill 1	Cut 1	Fill 2	Cut 2	Fill 3	Bridges	Fill 4	Cut 3	Fill 5	Top soil strip area	Fill 6	Cut 4	Tips Adj. to Fill 4	 East of Cut 4
Feature	Rock-fill embankment	Major cutting	Minor fill	Major cutting	Major fill, part constructed of rock-fill	Two bridges providing limitations to cross-haul	Removal of sand dunes below fill in part	Major cutting	Minor fill	No fill-topsoil strip only	Minor fill	Major cut-advanced earthworks		
Soil/ rock types		Boulder clay sand and gravel		Boulder clay			Blown sand	Boulder clay, clay, marl, conglomerate, limestone				Boulder clay, clay, clay, marl, sandstone, limestone		
Excavation techniques (soils)		Motor scrapers, dozers		Motor scrapers			Motor scrapers, tractor and box, dozers	Motor scrapers, dozers, back-acters				Draglines, face shovels, dozers, back-acters		
Excavation techniques (rock)								Drill and blast Dozers and rippers. Face-shovels, back-acters dropballs				Drill and blast, dozers and rippers, face shovels, back-acters dropballs		
Transport	Scrapers Scrapers Dump trucks Scrapers Dump trucks Imported rockfill-road lorries Scrapers Road lorries													

Table 10. Example of compaction plant used in practice *(Continued on facing page)*

Area	Fill type	Deposited by	Compacted by	Layer thickness, mm	Number of passes
ROUNDABOUT 1					
Slip roads	Thanet Sand	Lorries/dump trucks	140HP Crawler tractor towing vibrating roller (3260kg/m width)	275	8
Old chalk pits	Coarse sand-fine gravel (free draining)	Lorries	As above failed – then saturated material and compacted with tracks of 140HP crawler tractor	As required	As required
Landscaping	Chalk, clay, rubble	Lorries	140HP Crawler tractor towing vibrating roller (3260kg/m width)	200	2*
			As above without vibration on roller	150	2 to 4*
			140HP Crawler tractor towing 8000kg smooth wheeled roller	150	2 to 4*
MOTORWAY AND SIDE ROAD EMBANKMENT					
Start to Roundabout 1	Thanet Sand	Motor scrapers	300HP Crawler tractor towing 2 vibrating rollers in tandem (both 5600kg/m width)	100	1
Roundabout 1 to viaduct	Thanet Sand	Motor scrapers	300HP Crawler tractor towing 2 vibrating rollers in tandem (5600kg/m plus 3260kg/m)	100	1 – 2
			30740kg self propelled tamping roller	250	4
	Brickearth	Tractor & box	140HP Crawler tractor towing 5500kg towed tamping roller	225	8
Old chalk pits	Thanet Sand	Motor scrapers	140HP Crawler tractor towing 5600kg/m width vibrating roller	300	4
Surcharge to embankments	Thanet Sand	Dump trucks	140HP Crawler tractor towing 3260kg/m width vibrating roller	275	8

*Specification required compaction to be half of normal number of passes for landscaping fill

Area	Fill type	Deposited by	Compacted by	Layer thickness, mm	Number of passes
ROUNDABOUT 2					
Slip roads	Thanet Sand	Tractor & box	140HP Crawler tractor towing 5600kg/m width vibrating roller	300	4
		Dump trucks	170HP special rubber-tyred tractor towing 3260kg/m vibrating roller	275	8
	Gravelly clay	Tractor & box	140HP Crawler tractor towing 3260kg/m vibrating roller	200	4
		Motor scraper	30740kg self propelled tamping roller	225/150	4/12
Clay seal to contaminated area	London Clay	Lorries	140HP Crawler tractor towing 5500kg tamping roller	225	4
Trunk road embankments	Thanet Sand	Dump trucks	170HP special rubber-tyred tractor towing 3260kg/m vibrating roller	275	8
STRUCTURES					
Granular backfill	Sandy gravel	Lorries	140HP Crawler tractor towing 3260kg/m vibrating roller	200	4
			Manual 664kg/m vibrating roller	75	12

been spread over a number of embankments such that the rollers were required to be moved from area to area, it may have been more economical to use the three small capacity towed rollers.

Example on compaction practice

The example taken to illustrate compaction in practice concerns a recent motorway contract in south east England. The contract consisted of the construction of 2.25 km of motorway together with two roundabouts, their associated slip roads and a trunk road crossing the motorway. This involved the placing and compaction of over 700 000 m^3 of bulk fill, select backfill to structures, free drainage blankets and fill for surcharging embankments and for landscaping purposes. The compaction of the fill was carried out with a number of different types of plant which were selected to suit the materials being compacted, the volumes and rate of placement of fill requiring compaction, and the environment in which compaction was being carried out. A summary of the plant used is given in Table 10.

The majority of the fill used consisted of Thanet Sand. Where motor scrapers were employed to deliver the sand at high output rates, towed vibrating rollers in tandem or self-propelled tamping rollers were used, whereas when lorries, dump trucks or tractor and boxes were employed, single towed vibrating rollers were used. For other materials towed vibrating, smooth wheeled (deadweight) and towed or self-propelled tamping rollers were used, depending on the nature of the soil and rate of production required.

6. EARTHWORKS SAFETY

Accidents cause human anguish, physical suffering, loss of morale and a waste of resources. The amount of time lost arising from accidents in industry is greatly in excess of that lost through industrial disputes. Most accidents occur as a result of momentary carelessness or lack of proper appreciation of the task in hand.

Over the years a large number of regulations has grown up under successive Factories Acts. Every practising engineer and manager would do well to acquire a sound working knowledge of the construction laws which govern the working environment. In addition an employer has a common law 'duty of care' towards his employees, and a 'vicarious liability' for the actions of his servants. There are legal obligations to insure to cover the consequences of failing in these responsibilities. More recently, the Health and Safety at Work Act has converted what were mainly moral obligations into statutory duties. The greater a man's knowledge or authority, the more is expected of him.

Engineers who work in the construction industry should take steps to make themselves aware of the hazards which may exist there, so that they can plan their operations in such a way as to prevent them causing harm. When dealing with the great forces of nature or with fast moving heavy machinery, one is not always given a second chance. In the following paragraphs, some of the more common hazards that are encountered on sites involving earthworks are summarized.

Safety related to the construction of earthworks

(a) The failure of temporary slopes is a common cause of accidents and this may include failure of temporary batters in cuttings, embankments and trenches. An appraisal of the stability of such batters should be made if failure could cause a potential hazard and, if necessary, the batter cut to a safe angle or shoring constructed. Special care should be taken where groundwater is encountered as this will generally reduce the stability.

(b) Unless specifically allowed for in the design of the excavation, heavy plant should not be located, nor excavated material placed near the edge of the batters.

(c) Care should be taken when excavating from the base of a working face to ensure that the face is not overhanging or excessively high, so that if failure occurs there is no hazard to the driver, nor to the excavator.

(d) Where bench working is being carried out, care should be taken to ensure that the higher benches are not later undermined or that there is a risk of material falling or rolling from the upper slopes.

(e) Where excavations remain open overnight that could be a hazard to the public and site personnel; the excavations should be protected with barriers and wherever possible their presence should be indicated with hazard warning lights.

(f) Explosives should only be handled and used by experienced personnel and stored in accordance with the relevant regulations.

(g) When blasting is being carried out, a system should be operated to ensure that no-one approaches within a safe distance from the blast. If flying material occurs during a blast, the blasting techniques should be reviewed and the use of mats, etc. considered.

Safety related to earthmoving plant

(a) Heavy plant should be routed along distinct haul roads, preferably separate from pedestrians and light traffic.

(b) Haul roads should be kept in good condition as this will minimize the braking distance. Gradients should be kept to a minimum and in dry weather they should be watered to minimize the formation of dust clouds.

(c) Visibility from many large machines is poor and therefore it is advisable to keep well away from them. If this is difficult, such as when setting out or when in situ tests on the fill are being carried out, the drivers should be warned verbally or by roadside signs. Sitting in a vehicle may not be safe, as bulldozers can easily crush a Landrover of which the driver is unaware.

(d) Overhead obstructions such as overhead cables, bridge soffits, or temporary works that may cause clearance problems, should be indicated clearly by signs or tapes.

(e) Plant is generally less stable when travelling along a slope and should therefore be run up and down a slope whenever possible.

(f) Tipping on the embankments should be carried out short of the edge and then the fill dozed over the edge. If tipping is required at the edge of the tip then stop chocks should be provided.

(g) Vehicles that travel on public roads need by law to be in roadworthy condition. Where mud is spread on public roads by the wheels of site plant, the use of road cleaning equipment and wheel washers should be considered to minimize the possibility of accidents involving the public. This is especially important during the winter months.

ACKNOWLEDGEMENTS

The author is grateful for the assistance and comments provided by Messrs. A. A. Darby and D. Whitehouse of Bovis Civil Engineering Ltd., Messrs. W. S. Charles-Jones and T. G. Ley of John Laing Construction Ltd., Mr. H. G. Clapham of Ground Engineering Ltd. and Mr. M. Sarling of C. J. Prior (Earthmoving) Ltd. The Guide has been prepared with the support of John Laing Research and Development Ltd.

Table 8 is reproduced with the permission of the Controller of her Majesty's Stationery Office. The following companies have granted permission for the reproduction of photographs and tables and their assistance is acknowledged.

Anglo Swedish Equipment Ltd.
Atlas Copco (Great Britain) Ltd.
Bomag (Great Britain) Ltd.
Caterpillar Overseas S.A.
Hymac Ltd.
Hyster Europe Ltd.
H. Leverton and Co. Ltd.
Ransomes and Rapier Ltd.

BIBLIOGRAPHY

AKROYD T.N.W. *Laboratory testing in soil engineering.* Soil Mechanics Ltd., London, 1957.

ANON. Effective land clearing methods. *World Construction,*, Sept., 1979, pp 67 – 75.

ANON. Scrapers. *International Construction,* June, 1979, pp 65 – 77.

ARROWSMITH E.J. Roadworks fills – a material engineer's viewpoint. *Clay Fills*, ICE, London, 1979.

ATLAS COPCO. Atlas Copco Manual. 3rd edition, Sweden, 1978.

BEAMAN A.L. and JONES P.D. *Noise from construction and demolition sites measured levels and their prediction.* Construction Industry Research and Information Association (CIRIA) Report No. 64, 1977.

BISHOP A.W. The use of the slip circle in the stability analysis of slopes. *Geotechnique*, 1955, Vol. 5, pp 7 – 17.

BISHOP A.W. and MORGERNSTERN N.R. Stability coefficients for earth slopes. *Geotechnique*, 1960, Vol. 10, pp 147 – 192.

BLYTH F.G.H. and DE FREITAS M.H. *A geology for engineers.* 6th edition, Edward Arnold, London, 1974.

BRITISH STANDARDS INSTITUTION. *Methods of testing soils for civil engineering purposes*. BS 1377, 1975.

BRITISH STANDARDS INSTITUTION CP 2001. *Site investigations*, 1957, (under review).

BRITISH STANDARDS INSTITUTION CP 2003. *Earthworks*, 1959.

CARSON A.B. *General excavation methods*. F.W. Dodge Corpn., USA, 1961.

CATERPILLAR TRACTOR COMPANY. *Performance handbook*. 10th edition, 1979.

CHANDLER R.J. and SKEMPTON A.W. The design of permanent cutting slopes in stiff fissured clays. *Geotechnique*, 1974, Vol. 24, pp 457 – 466.

COLLING J.G. *Forecasting trafficability of soils. Report 10. Relations of strength to other properties of fine grained soils and sands with fines*. Tech. Memo No. 3-331, US Waterways Experimental Station, Vicksburg, Mississippi, 1971.

DAY D.A. *Construction equipment guide*. Wiley, New York, 1973.

DENNEHY J.P. The remoulded undrained shear strength of cohesive soils and its influence on the suitability of embankment fill. *Clay Fills*, Institution of Civil Engineers, London, 1979.

DEPARTMENT OF THE ENVIRONMENT. *Specification for road and bridge works*. HMSO, London, 1976.

DEPARTMENT OF THE ENVIRONMENT. *Notes for guidance on the specification for road and bridge works*. HMSO, London, 1976.

DEPARTMENT OF THE ENVIRONMENT. *Specification for road and bridge works*. Supplement No. 1, HMSO, London, 1978.

DREIBLATT D. *The economics of heavy earthmoving*. Praeger, New York, 1972.

DUMBLETON M.J. and WEST G. *Preliminary sources of information for site investigations in Britain*. Dept. of the Environment, Transport and Road Research Laboratory, Laboratory Report 403 (revised edition), Crowthorne, 1976.

DUNCAN N. *Engineering geology and rock mechanics*. Leonard Hill, 1969.

EDMUNDS P. Rock drilling in perspective. (2 parts), *Civil Engineering* Sept/Oct, 1979.

FARRAR D.M. and DARLEY P. *The operation of earthmoving plant on wet fill*. TRRL, Report 688, Crowthorne, 1975.

FISCHER F. *The technical aspects of compaction in earthmoving and road construction*, 4th edition. Clark International Marketing, SA, 1975.

FORDE M.C. and DAVIS F.G. Trafficability studies on wet clay. *Clay Fills*, Institution of Civil Engineers, London, 1979.

HAMMOND R. *Earthmoving and excavating plant*. C R Books, London, 1964.

HAVERS J.A. and STUBBS F.W. *Handbook of heavy construction*. 2nd edition. McGraw-Hill, New York, 1971.

HOEK E. and BRAY J.W. *Rock slope engineering*. Institution of Mining and Metallurgy, London, 1975.

HYSTER COMPANY. *Compaction handbook*. Sixth edition. Illinois, 1978.

INGERSOLL RAND COMPANY. *Compaction data handbook*. Shippensburg, 1979.

KENNARD M.F. et al. Shear strength specifications for clay fills. *Clay fills*. Institution of Civil Engineers, London, 1979.

MALCOLM T. and KIMBER R.W. Selection, procurement and management of construction equipment. *Inst. Civil Engineers Reference Book*, Ed. L.S. Blake, 3rd edition, Butterworths, London, 1975.

MANTON B.G. *Compaction. Present trends and modern plant reviewed*. Current Construction Practice No. 2, Building and Contract Journals Ltd., London, 1966.

MEAD H.T. and MITCHELL G.L. *Plant hire for building and construction*. Newnes-Butterworth, London, 1972.

NICHOLS H.L. *Moving the earth*. North Castle Books, 1962.

NORMAN R. *The effect of wet weather on the construction of earthworks*. CERA (CIRIA), Research Publication No. 3, 1974.

PARSONS A.W. *The efficiency of operation of earthmoving plant on road construction sites*. TRRL, Supplementary Report 351, Crowthorne, 1977.

PARSONS A.W. *The rapid measurement of the moisture condition of earthwork material*. TRRL, Laboratory Report 750, Crowthorne, 1976.

PEURIFOY R.L. *Construction planning, equipment and methods.* McGraw-Hill, New York, 1970.

READ C.P. Compactors. *Construction Industry International*, July, 1979, pp 36 – 55.

RODIN S. Earthworks – some practical aspects in the UK. *Muckshifter and Bulkhandler*, Vol. 22, 1964.

SCHROEDER W.L. *Soils in construction*. Wiley, New York, 1975.

SKEMPTON A.W. Long term stability of clay slopes. *Geotechnique*, 1964, Vol. 14, pp 77 – 101.

SNEDEKER E.A. Choice of an upper limit of moisture content for highway earthworks. *Highways Design and Construction*, January, 1973.

STEPHENS J.H. *The Guinness book of structures,* 1976.

STONE N.J. Geotechnical aspects of estimating and execution of dredging contracts. *International Dredging and Port Construction*, January, 1979, pp 15 – 21.

TERZAGHI K. and PECK A.B. *Soil mechanics in engineering practice*. Wiley, New York, 1967.

THOMSON G.H. and PEARCE R.W. Ground treatment of fill. *Consulting Engineer*, (2 parts), Feb/March, 1979.

TRANSPORT AND ROAD RESEARCH LABORATORY. *Soil mechanics for road engineers*, HMSO, London, 1952.

UNITED STATES DEPT. OF THE INTERIOR BUREAU OF RECLAMATION. *Earth manual*, 1963.

VAUGHAN P.R. et al. Factors controlling the stability of clay fills in Britain. *Clay Fills*. Institution of Civil Engineers, London, 1979.